YOUR KNOWLEDGE HAS VALUE

- We will publish your bachelor's and master's thesis, essays and papers

- Your own eBook and book - sold worldwide in all relevant shops

- Earn money with each sale

Upload your text at www.GRIN.com
and publish for free

Vijay Kothari, Haren Gosai, Shreya Raval, Vimla Chaudhary

Natural products as potential sources of inhibitors of bacterial quorum-sensing

GRIN Publishing

Bibliographic information published by the German National Library:

The German National Library lists this publication in the National Bibliography; detailed bibliographic data are available on the Internet at http://dnb.dnb.de .

Imprint:

Copyright © 2014 GRIN Verlag GmbH
Print and binding: Books on Demand GmbH, Norderstedt Germany
ISBN: 978-3-656-87121-7

This book at GRIN:

http://www.grin.com/en/e-book/286836/natural-products-as-potential-sources-of-inhibitors-of-bacterial-quorum-sensing

Natural products as potential sources of inhibitors of bacterial quorum-sensing

Vijay Kothari

Haren Gosai

Shreya Raval

Vimla Chaudhary

Acknowledgement

Authors thank Nirma Education and Research Foundation (NERF), Ahmedabad for financial and infrastructural support. All authors thank their respective family members for their unconditional support.

1. INTRODUCTION

Nature has been a source of medicinal agents for thousands of years since it comprises of compounds that are highly diverse and often provide highly specific biological activities. This follows from the proposition that essentially all natural products have some receptor binding capacity (Verdine GL., 1996).To promote the proper use of herbal medicine and to determine their potential as sources for new drugs, it is essential to study medicinal plants, which have folklore reputation in a more intensified way(Awadh NA *et al.*, 2001).

Some natural products have been approved as new antimicrobial drugs, but there is a continuous and urgent need to screen more and more plant species and discover new antimicrobial compounds with diverse chemical structures and novel mechanisms of action to combat new and re-emerging infectious diseases of today's era(Rojas R *et al.*,2003). Plant based natural constituents can be derived from any part of the plant like bark, leaves, lowers, roots, fruits, seeds, etc(Gordon MC *et al.*, 2001).i.e. any part of the plant may contain active components. World Health Organization has also approved the study of medicinal plants for the development of new drug lead (WHO, 2000).

There is an evolving group of chronic infectious diseases and condition in which bacteria is present and they are difficult to culture and demonstrate consistently hence resistant to current antimicrobial and tools and often require surgical removal to resolve. Despite adequate drainage antibiotics are not the **"silver bullet"** once envisaged for this chronic infectious condition (Hayes GW *et at.*, 1993). The use of antimicrobials from natural sources could have been a great impact for preserving food storage from contamination, and in controlling plant and human diseases of microbial origin (Balandrin *et al.*, 1985).The continued evolution of infectious disease and the resistance offered by the pathogens to the existing pharmaceuticals have led to an intensified search for the new availability of antimicrobials against fungal, bacterial and viral particles as plants constantly interact with the rapidly changing and damaging environmental conditions. Being organism devoid of mobility plants has developed a strong immunity against the metabolites and chemical stress which come across their life. The ability of plant to carry out combinatorial chemistry by mixing matching and evolving the gene products required for certain secondary metabolite biosynthetic pathway create an unlimited pool of chemical compound, which humans have exploited to their benefit.

Many pathogenic bacteria trigger the production of their virulence factors in a population density-dependent manner, a cell-to-cell communication mechanism known as **quorum sensing** (Vandeputte *et al.*, 2009).Quorum sensing (QS) is mediated through small signal molecules called auto inducers that regulate the target gene expression responsible for the phenotypes essential to pathogenicity/symbiosis. In Gram negative bacteria, QS is mediated through N-acyl homoserine lactones (AHL) (Bassler BL., 2002).

QS has led to its role in controlling the production of virulence factors such as exopolysaccharide synthesis, biofilm formation, swarming motility, pigment production and antibiotic production in some pathogenic bacteria (Smith JL *et al.*,2011, Schonewille E *et al.*, 2012). Increasing emergence of antibiotic resistance in Gram negative bacteria also demands alternative strategies to combat bacterial infections and anti-QS approach is being viewed as an attractive alternative.

Plant metabolites having the ability to control the growth of microbes have been traditionally used to treat human diseases including microbial infections (Shahid M *et al.*, 2009). There are certain plants such as carrot, garlic, habanero (chilli), and water lily produce compounds that interfere with bacterial QS. Other plants found to be having anti-quorum sensing potential were pea (*Pisum satavium*) seedlings, crown vetch, soybean, and tomato (Rasmussen *et al.*, 2006). Plants have recognised symbiotic relationship with colonizing bacteria. There may be plant based natural quorum sensing signals to control pathogens. It was observed that certain seaweed plants never became cover with bacteria or higher organism (called bio fouling, which is also observed on sub surfaces of ships) because the plant produced halogenated furanones, which have quorum sensing inhibitors activity(QSIs), but these halogenated furanones contain halogen which make them restricted to human use. Therefore, large scale screening of QSIs in nature was therefore performed. In recent years multiple drug resistance occur against commercially available antimicrobials which has forced scientist to search for new antimicrobials from plants which has vast source of bioactive agents in it.

➢ **Plant materials:**

1. Pongamia pinnata

Family- Fabaceace

Common name- karanj, and is also called as Derris indica in India and Myanmar.

A.Tree   B. seeds

Fig 1 – *Pongamia pinnata*: A. growing in natural habitat; B. Matured seeds.

Pongamia pinnata (Syn. *Milleta pinnata*) is a species of tree in the pea family. It is native in tropical and temperate Asia including parts of India, China, Japan, Malaysia, and Australia. *Pongamia pinnata* also called as Derris indica is a monotypic genus and grows abundantly along the coasts and riverbanks in India and Myanmar. The tree is known for its multipurpose benefits and as potential sources of biodiesel (Kesari V *et al.*, 2009, Naik M *et al.*, 2008). The seeds are reported to contain on an average about 28–34% oil with high percentage of polyunsaturated fatty acids. Historically, Pongamia has been used as folk medicinal plant, particularly in Ayurvedha and Siddha systems of Indian medicine (Meera B *et al.*, 2003). All parts of the plant have been used as a crude drug for the treatment of tumours, piles, skin diseases, itches, abscess, painful rheumatic joints wounds, ulcers, diarrhea etc.

Besides, it is well known for its application as animal fodder, green manure, timber and fish poison.
It recognized to possess applications in agriculture and environmental management, with insecticidal and nematicidal activity. More recently, the effectiveness of *P. pinnata* as a source of biomedicines has been reported (Brijesh S *et al.,* 2006), specifically as antimicrobial and therapeutic agents.

2. *Manilkara hexandra*

Family- Sapotaceae

Common name- rayan (Hindi), raina, khirni(Marathi), Ceylon Iron Wood

a.fruit

b.seeds

(Fig 2 –*Manilkara hexandra*)

Manilkara hexandra(Roxb.) Dubard belonging to family Sapotaceae, is a socio-economically important underutilized fruit species of western-central India. It is locally known as 'Khirni', 'Raina' and 'Rayan' by tribal people of different states of India. It is believed to be originated in India. The tree is found as natural wild in the south, north and central India mostly in the states of Rajasthan, Gujarat, Madhya Pradesh and Maharashtra. It is a commercially and medicinally important tropical tree species and is a significant source of livelihood and nutritional support species for local tribal population. In India, this species is occasionally cultivated in backyards, homestead gardens, public parks as avenue tree and in farmer's fields near villages due to its economic importance as fruit tree having nutritional and medicinal properties.

Fruits of Khirni have high economic value as mature fresh fruits which are sweet and a good source of minerals, sugars, protein, carbohydrate and vitamin A (Pareek J *et al.* 1998). Whole partially matured fruits are dried in the sunlight to reduce the chances of infection and deterioration. Tribal people sell the fresh as well as dried fruits in the local market at the cost of Rs. 30–40 per kg. Each tree provides fruits worth of Rs. 500–2,000/- to a tribal family, which is a substantial livelihood support to them. Fresh fruits are generally consumed by entire tribal family which provides good nutritional support in the form of vitamin A, hence whole fruit works as **"vitamin A capsule"** for the nutritionally deficient tribal population especially children and women. Bark and fruits are also used for several medicinal purposes like treatment of ulcers, dyspepsia, opacity of the cornea, bronchitis, urethrorrhea, leprosy, etc. (Pareek *et al.* 1998;Chanda and Parekh 2010).

The seeds contain approximately 25 % oil, which is used for cooking purposes. The bark also contains 10 % tannin, which is used for treatment of fever and may be utilized in tanning purposes. The wood of this tree is very hard, tough and durable and is used for oil presses, house building and turnery.

3. *Pyruspyrifolia*

Family-Rosaceae

Common name- nashpati, pear, Chinese pear, apple pear, sand pear

a. Fruits   b.seeds

(Fig 3 – *pyrus pyrifolia* (http://www.floridata.com/ref/p/pyrus_pyrifolia.cfm)

The sand pear is larger than most kinds of pear trees, reaching as high as 40 ft. (12 m), with a rounded crown that may spread 20 feet (6 m) or more across. The tree is almost completely covered with white flowers, putting on a spectacular show in early spring. *Pyrus pyrifolia* is native to China and Japan. Dozens of cultivars are grown in Asia. The fruits are hard and the flesh is grainy, some say "sandy" in texture. They are most useful for making pear butter, preserves, pies, and for canning. Pears are a good source of dietary fiber and vitamin C. Pear wood is one of the preferred materials in the manufacture of high-quality woodwind instruments and furniture. pears are used to treat nausea.

➢ **Test microorganisms**

- ***Chromobacterium violaceum*** (MTCC:2656)

Chromobacterium violaceum is a Gram-negative rod which is found in the soil and water of tropical and subtropical areas. *Chromobacterium violaceum* (*Cv*) appears to be an opportunistic pathogenic bacterium, which affects humans and animals in subtropical and tropical areas. The low infectious capability of *Cv* is evidenced by the fact that rivers such as Negro River, in the Amazon region of Brazil, where the bacteria is highly abundant, are sources of drinking water, without the occurrence of widespread infection among the local population(Groves M *et al,*. 1969).

2. MATERIALS AND METHODS

> **Plant material**

The plants were selected on the basis of earlier reported antimicrobial activity of their other parts (pulp, stem, leaf, fruits, etc.), (Arote SR et al., 2009; Kesari V et al., 2009), and same parts against other organisms. The seeds of three plants *Pongamia pinnata (Karanj), Pyrus Pyrifolia (Nashpati), Manilkara hexandra (Rayan)* were collected during months of October, 2013 to November, 2013. They were authenticated for their unambiguous identity by Dr.Himanshu Pandya, Department of Botany, Gujarat University, Ahmedabad. They were thoroughly washed with tap water, air-dried in shade and stored in opaque air-tight containers (at room temperature) to avoid photo-oxidation(Houghton and Raman, 1998). The seeds were checked at regular intervals for any physical or biological damage. Damaged seeds were removed from the collection.

Table1. Details of plant seeds selected

Scientific Name	Image (seeds)	Family	Common Name	Parts reported for antimicrobial Activity
P. pinnata		Fabaceace	Karanj	Leaves , Flower (Arote SR *et al.*, 2009;Kesari V et al., 2009)
P. pyrifolia		Rosaceae	Nashpati	Fruit (Cho Y *et al.*, 2013)

M. hexandra		Sapotaceae	Rayan	Leaves, fruit (Pareek *et al.* 1998)

> ## ➢ Test organisms:

 Following test organisms were procured from Microbial Type Culture Collection (MTCC), Chandigarh.

Table 2.Test organisms

No.	Organism	MTCC Code	Medium	#Remarks
1	*C. violaceum*	2656	Nutrient agar	Sensitive to streptomycin, Gentamicin till 10µg/mL, Resistance toCephazolin till 100µg/mL

#Antibiotic susceptibility pattern determined by macro broth dilution assay in our lab

➢ Microwave Assisted Extraction (MAE):

Seeds were extracted by MAE method earlier published by us (Kothari et al., 2009, Kothari et al., 2011; Ramanuj et al., 2012; Darji *et al.*, 2012). Seeds were crushed twice in grinder (Maharaja whiteline bonus grinder) to coarse powder at knob 1 for 60 sec. One gram of seed powder was soaked into 50 ml of respective solvent in brown coloured screw capped bottle (250 ml, Merck, Mumbai, India), cap was loosened slightly to avoid pressure build up. Solvents used for extraction were acetone, methanol (Merck, Mumbai) and ethanol (50%; Ureca consumers Co. Op Stores Ltd., Ahmedabad India). Then this was subjected to microwave assisted extraction (MAE). For this, bottle was kept in microwave oven (Electrolux, EM30EC90SS) for extraction at 720W. At a time only one bottle was kept for procedure. The time for extraction of all seeds for respective solvent is reported in Table 3. The extraction was followed by macrofiltration (nylon strainer), which was further subjected to centrifugation at 7,500 rpm for 20 min. Then after microfiltration was carried out using Whitman filter paper no. 1 (Whatman International Ltd., England) to remove the fine particulate matter and was allowed to evaporate in pre-weighed petridishes (in weighing balance, Setra, BL-410S) a temperature below the boiling point of respective solvents. For determination of antimicrobial activity, extract were reconstituted in dimethyl Sulfoxide (DMSO, Merck, Mumbai) as it possesses the ability to dissolve polar as well as non-polar compounds. Extract was collected in sterile flat bottom glass vials (15ml, Merck, Mumbai) which were protected from light to avoid photo-oxidation of light sensitive compounds (Houghton and Raman, 1998). The internal surface of vial cap was wrapped with aluminium foil to avoid leaching of vial cap material and their absorption by extract (Houghton and Raman, 1998). They were then stored in refrigerator at 4°C.

Table 3. Heating and cooling cycles of different Solvents during MAE are as follow:

Solvents	Total extraction time	Total heating time (s)	First heating time (s)	Cooling Time (s)	Reheating Time (s)
Ethanol (50%)	8 min 6 s	120	25	40	10
Methanol	12 min 16 s	90	10	40	5
Acetone	3 min 6 s	70	40	40	10

Extraction efficiency and reconstitution efficiency was calculated by using following formula:

- **Extraction efficiency:**

$$\text{Extraction efficiency} = \frac{\text{weight extracted (mg)}}{\text{weight of initial material (mg)}} \times 100$$

Weight extracted (mg) = (Weight of petriplate after evaporation of solvent) - (weight of empty plate)

- **Reconstitution efficiency:**

$$\text{Reconstitution efficiency} = \frac{\text{weight of extracted reconstituents (mg)}}{\text{total weight of dried material (mg)}} \times 100$$

Weight of extract reconstituted (mg) =

(Weight of petriplate after evaporation of solvent) -   (weight of petriplate after reconstitution)

➤ **Inoculum standardization:**

For Bacteria:

5ml sterile normal saline was taken in sterile test tube in which few (3-4) isolated colonies of 24h old culture were added and turbidity was visually matched with 0.5 McFarland standard. This 0.5 McFarland turbidity standard at 625 nm is equivalent to1.5 × 10^8 CFU/ml.

➤ **Antimicrobial   susceptibility   testing   (AST)   with   planktonic cells (Macrobroth Dilution method):**

The test organism was challenged with different concentrations of *M. hexandra,* *P. pyrifolia, P. Pinnata* seed extracts. These seed extracts were tested only against *C. violaceum.* Nutrient broth (Himedia) was used for AST, because MH and minimal media did not support the solubility of pigment of *C. violaceum.* Total volume of the assay system was kept 2ml. Extracts of *P. pyrifolia* were getting precipitated in N broth. The inoculum was prepared in normal saline solution (0.85 % NaCl) from 24 h old culture of bacteria. Inoculum density of the test organisms was adjusted to that of 0.5 McFarland standards. Various controls and experimental tubes were prepared as below and the final volume in each tube was 2ml. In all controls (except sterility and turbidity) 10%v/v inoculum was added. Turbidity controls (abiotic control) were kept to nullify the contribution of extract itself towards total turbidity. As Positive control gentamicin and streptomycin were used at 10 µg/mL.

- Sterility Control: 2ml uninoculated sterile Nutrient broth.
- Growth Control: 1.8 ml N broth + 200 µl inoculum.
- Negative Control: 1.78 ml N broth + 20 µl DMSO + 200 µl inoculum.
- Positive Control: 1.78 ml N broth + 20 µl antibiotic + 200 µl inoculum.
- Turbidity Control: 1.78 ml N broth + 20 µl extract + 200 µl normal saline.
- Experimental: 178 µl broth + 2 µl extract + 20 µl inoculum.

Normally 1% v/v extract was added in the well, but in case where the concentration of available extract was low, 2% or 3% v/v was added to achieve higher concentration in tube. Respective to it negative control also changed. Here streptomycin and Gentamicin (Himedia) (10µg) were taken as positive control. After inoculation incubation was carried out at 35±0.5ºC for 18-20 h. After incubation there was visual   observation of turbidity as well as violacein production by organism against extract was also visually observed. The optical density measurement was taken in spectrophotometer (Elico Thermoscientific, Spectrometric 20 D+) at 660 nm. Before taking optical density vortexing of tubes was done for 5 sec to create uniform condition in tubes. After completion of taking optical density at 660 nm there was violacein extraction procedure was performed. The optical density measurement of extracted violacein was taken at 585 nm.

❖ Quantification of Violacein[Choo et al., 2005]:

- ➢ Growth of cells were measured at 625 nm (Thermoscientific, Spectrometric 20 D+). Before measurement of growth all samples were vortexed for 10 seconds.

- ➢ Then 2 mL of sample was transferred into eppendoff vials (2 mL).

- ➢ Centrifugation was done at 10000 rpm for 15 min at 25 ˚C to pellet down cells. (Eppendorf 5417R)

- ➢ After centrifugation cell free supernatant was discarded.

- ➢ 2 mL of Dimethysulfoxide (DMSO) were added in each pellet to resuspend the pellet.

- ➢ Vortexing was done to resuspend the pellet till pellet was dissolved fully.

- ➢ Centrifugation was done to extract violacein at 10000 rpm for 15 min at 25 C.

- ➢ After centrifugation absorbance was measured at 585 nm to estimate violacein.

- ➢ **Prodigiosin spectrum:**

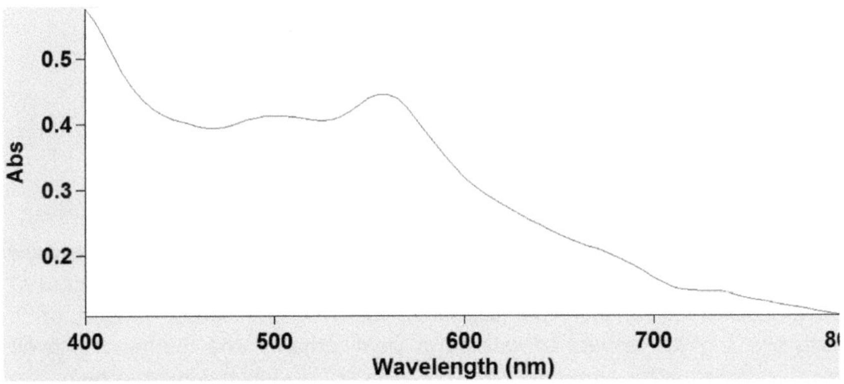

Prodigiosin was absorb at 560 nm(λ_{max})

3. Results

1. The extraction efficiency and reconstitution efficiency of various extracts of all the seeds are shown in Table1.

Table 1: Extraction and reconstitution efficiency for all the seed Extracts.

Seed	Solvent	Extraction Efficiency (%)	Reconstitution Efficiency (%)
P. pinnata	Ethanol (50%)	**21.58**	19.28
	Methanol	17.08	29.86
	Acetone	17.42	26.41
M. hexandra	Ethanol (50%)	13.24	**83.23**
	Methanol	13.78	74.31
	Acetone	12.00	41.66
P. pyrifolia	Ethanol (50%)	10.40	78.85
	Methanol	9.78	77.09
	Acetone	9.32	17.59

Extraction and reconstitution efficiency of all the seeds in different solvents is recorded in Table1 Highest extraction efficiency (21.58%) was obtained in case of hydro alcoholic extract of *P. pinnata*seeds, followed by acetone extract of *P. pinnata* seeds. With respect to extraction yield, ethanol and methanol proved better than acetone for all seeds. Polar solvents are usually believed to be better for MAE than the non-polar ones because they have high dielectric constants (Proestos C and Komaitis M, 2007). Hydroalcoholic extract of *M. hexandra* seeds and *P. Pyrifolia* seeds were giving relatively higher extraction efficiency (compared to other solvents, Table 1) which indicates that these experimental seeds contain components which are soluble in polar solvent like ethanol. This result suggests that ethanol is most suitable solvent for this type of extraction purpose. It has been also suggested that addition of some amount of water in solvent enhances the extraction efficiency. This may be due to increase in swelling of plant material by water thereby increasing the contact surface area between the plant matrix and solvent (Mandal et al., 2007). It can be inferred that methanolic extract of *P. pinnata*seeds was giving higher extraction efficiency (29.86%) compared to its methanol and acetone extracts (Table 1). Extraction efficiencies of ethanolic extract of *M. hexandra, P. Pinnata* and *P. Pyrifolia* seeds are higher than methanolic extract of same seeds, which shows that seed contain certain components which are more soluble in ethanol than methanol. Extraction efficiency of methanolic extracts of *M. hexandra* is higher than ethanolic extract of same seed (Table 1), which indicate that these seeds contain certain components which are more soluble in methanol than ethanol. These all results suggest that extraction efficiency depend on type of component present in the plant material. From Table 1, it is evident that extraction efficiency of acetone extract is low as compared to other solvents. Reconstitution efficiency is the % amount of material recovered form 1g of seed extract. The reconstitution efficiency is higher in *M. hexandra* and *P. Pyrifolia* seeds in case of ethanol extract (compared to other solvent, Table 1), so it can be concluded that ethanol is a better solvent with respect to extraction as well as reconstitution for these seeds. While reconstituting the dried extracts of *P. pinnata*in DMSO, higher reconstitution efficiency was obtained with their acetone extracts. It should be noted that the possible advantage(s) of high extraction yield may be somewhat compromised by low reconstitution efficiency, because some of the phytoconstituents present in the original extract may be left out due to inability of the reconstituting solvent to solubilize all of them.

2. **Anti-quorum sensing activities in *P. Pinnata* plant using Antimicrobial susceptibility testing of organism**

Loss of purple pigment in *Chromobacterium violaceum*is indicative of QS inhibition by the plant products. The anti-QS activity of active species as fresh, dried, ethanol-extract, methanol-extract, aceton extract were screened using macro broth dilution method. Control containing   DMSO was included. The best QS inhibition was seen with ethanolic and methanolic extract of *P. pinnata* and a minor QS inhibition was apparent with aceton extract of same seed.

Table: 2. Anti-quorum sensing activity of Ethanolic extract of karanj

Conc. (µg/mL)	Growth (OD_{660}) (Mean±SD)		% Inhibition compared to control	Violacein (OD_{585}) (Mean±SD)		% Inhibition compared to control	Volacein Unit (OD_{585}/OD_{660})		%change compared to control
	C	E		C	E		C	E	
250	1.73±0.01	1.15±0.01	33.52[**]	0.67±0.05	0.30±0.00	55.22[*]	0.38	0.26	-31.57
500	1.57±0.08	1.04±0.01	33.75[*]	0.41±0.02	0.12±0.00	70.73[**]	0.26	0.11	-57.69
750	1.57±0.08	0.90±0.03	42.67[**]	0.41±0.02	0.10±0.00	75.60[**]	0.26	0.11	-57.69
1000	1.57±0.08	0.93±0.00	40.76[**]	0.41±0.02	0.07±0.00	82.92[**]	0.26	0.07	-73.07

[*]$p<0.05$; [**]$p<0.01$

Ethanolic extract of Karanj seeds shown the highest anti-quorum sensing activity in our study. At concentration of 1000 µg/ml quorum sensing was inhibited by around 82.92% followed by 75.60% decrease at 750 µg/ml concentrations. Violacein gave dose dependent response at significant level. As concentration was increase relatively violacein inhibition was also increased.

> ### HPLC of ethanolic extract of karanj

For reverse phase HPLC (Agilent 1200 infinity) 5µl of ethanolic extract of karanj (reconstituted in DMSO) was injected into phosphoric acid: water (95:5) equilibrated semi-preparative C_{18} column (Eclipse plus C_{18}-Agilent). The column was eluted for 35 min in phosphoric acid: water followed for 35 min with a linear phosphoric acid: acetonitrile (50:50) gradient as mobile phase at the flow rate of 1mL/min. Absorbance was monitored at 270 nm.

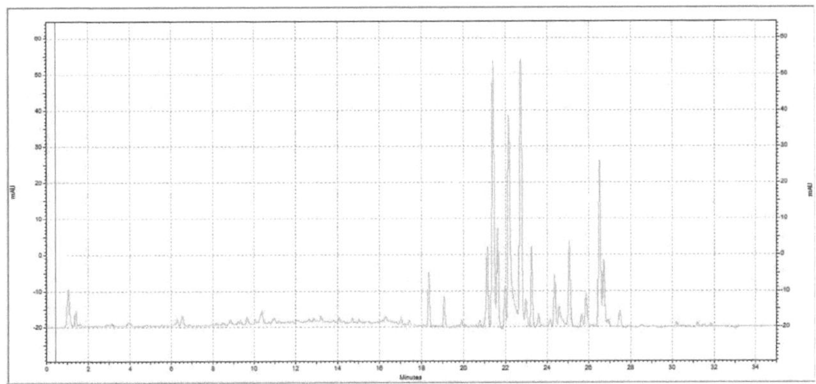

Fig.1. Reverse phase HPLC ofEthanolic extract of karanj

In Figure 1 the compounds are represented by different peaks separated in time in the chromatogram. Each elutes at a specific location, measured by the elapsed time between the moment of injection [time zero] and the time when the peak maximum elutes. By comparing each peak's retention time [t_R] with that of injected reference standards in the same chromatographic system [same mobile and stationary phase], It may be able to identify each compound. Here the polar compounds were eluted first and then after nonpolar compounds were eluted, and the nonpolar compound has the best resolution peak, so we can simply said that more than one nonpolar compound which were present in ethanolic extract of karanj, they were responsible for the anti-quorum sensing effect against *C. violaceum.*

1

Table: 3. Anti-quorum sensing activity by Methanolic extract of Karanj

Conc. (µg/mL)	Growth (OD_{660}) (Mean±SD)		%change compared to control	Violacein (OD_{585}) (Mean±SD)		% change compared to control	Volacein Unit (OD_{585}/OD_{660})		%change compared to control
	C	E		C	E		C	E	
250	1.21±0.05	1.46±0.10	20.66↑	0.28±0.00	0.42±0.03	50↑	0.23	0.28	21.73↑
500	1.21±0.05	0.93±0.00	-23.14[*]	0.28±0.00	0.17±0.00	-39.28[**]	0.23	0.18	-21.73
750	1.40±0.01	0.86±0.01	-38.57[**]	0.25±0.00	0.15±0.01	-40.00[**]	0.17	0.17	0
1000	1.56±0.01	0.83±0.03	-46.79[**]	0.46±0.00	0.11±0.00	-76.08[**]	0.29	0.13	-55.17

[*] $p<0.05$; [**] $p<0.01$

Methanolic extract of Karanj seeds shown best anti quorum sensing activity after ethanolic extract of same seed in our study. At concentration of 1000 µg/ml quorum sensing was inhibited by around 76.08% followed by 39.28% decrease at 500 µg/ml concentrations. But at the low concentration it enhanced quorumsensing.

Table: 4. Anti-Quorum sensing activity by acetone extract of karanj

Conc. (µg/mL)	Growth (OD_{660}) (Mean±SD)		%inhibition compared to control	Violacein (OD_{585}) (Mean±SD)		%inhibition compared to control	Volacein Unit (OD_{585}/OD_{660})		%change compared to control
	C	E		C	E		C	E	
250		1.28±0.01	13.51		0.38±0.01	20.83		0.29	-9.37
500	1.48±0.11	1.31±0.01	11.48	0.48±0.05	0.29±0.01	39.58[*]	0.32	0.22	-31.25
750		1.45±0.06	2.02		0.47±0.08	2.08		0.32	0
1000		1.37±0.05	7.43		0.46±0.05	4.16		0.33	3.12↑

[*] $p<0.05$

Here, from the above table it can be seen that acetone extract of Karanj was found giving different effect when tested at different concentrations. Quorum sensing was inhibited in *Chromobacteriumviolaceum*but at very low level. Extract showed highest anti-quorum sensing activity at the concentration of 500 µg/ml followed by 250 µg/ml.

3. Anti-quorum sensing activities of prodigiosin

*Serratia*strains produce prodigiosin.Prodigiosin is typical secondary metabolite only appearingin the later stages of bacterial growth Prodigiosin is of potential clinical interest because it isreported to have anti-fungal, anti-bacterial, anti-protozoal/anti-malarial, immunosuppressive and anti-cancer activities(Williams and Quadri, 1980; Tsuji *et al.*, 1992; Demain, 1995;D'Alessio*et al.*, 2000;Montaner*et al.*, 2000)

Table: 5. Anti-quorum sensing activity of prodigiosin

Conc. (µg/mL)	Growth (OD_{660}) (Mean±SD)		%Inhibition compared to control	Violacein (OD_{585}) (Mean±SD)		%Inhibition compared to control	Volacein Unit (OD_{585}/OD_{660})		%change compared to control
	C	E		C	E		C	E	
38.8	0.94±0.01	0.73±0.01	22.34[**]	0.10±0.00	0.08±0.01	20.00	0.10	0.10	0
77.6	0.92±0.01	0.79±0.00	14.13[**]	0.09±0.01	0.06±0.00	33.33	0.09	0.07	-22.22
116.4	0.89±0.01	0.79±0.00	11.23[*]	0.11±0.00	0.06±0.00	45.45[**]	0.12	0.07	-41.66

[*]$p<0.05$; [**]$p<0.01$

Here the concentration of the prodigiosin was 38.8µg/mL at 1%v/v and as we went upto 2%v/v and 3% the concentration were 77.6µg/mL and 116.4µg/mL. Because of the lesser content of prodigiosin we kept the concentration upto 3%v/v for quorum sensing experiment.

Prodigiosin is apink colored pigment produced by *S.marcescens* when extracted out and tested against *Chromobacteriumviolaceum* showed an antiquorum sensing activity by remarkable decrease in per cell violacein production when tested at higher concentrations (116.4µg/ml).

4. Quorum sensing enhancing activities in *M. hexandra* and *P. pyrifolia* plants using Antimicrobial susceptibility testing of organism

The violet pigment violacein is an indole derivative, isolated mainly from bacteria of the genus *Chromobacterium*, which exhibits important antitumoural, antimicrobial and antiparasitary properties. Furthermore, the formulation of violacein in different polymeric carriers developed so far offers alternative approaches to overcoming physiological barriers and undesirable physicochemical properties *in vivo*, thus improving its efficacy. (Durán.,*et al* 2010)

Table: 6. Quorum sensing enhancing byEthanolic extract of rayan

Conc. (µg/mL)	Growth (OD_{660}) (Mean±SD)		%change compared to control	Violacein (OD_{585}) (Mean±SD)		%Change compared to control	Volacein Unit (OD_{585}/OD_{660})		%change compared to control
	C	E		C	E		C	E	
250	1.31±0.00	1.28±0.00	-2.2[*]	0.13±0.01	0.24±0.00	84.61[**]↑	0.09	0.18	100↑
500	1.31±0.00	1.74±0.00	32.82[**]↑	0.13±0.01	0.70±0.00	438.46[**]↑	0.09	0.40	344.44↑
750	1.31±0.00	1.80±0.02	37.40[**]↑	0.15±0.00	0.68±0.00	353.33[**]↑	0.11	0.37	236.36↑
1000	1.27±0.00	1.89±0.10	48.81[*]↑	0.27±0.00	0.64±0.00	137.03[**]↑	0.21	0.33	57.14↑

[*]$p<0.05$; [**]$p<0.01$

Ethanolic extract of Rayan seeds was observed to be having quorum sensing enhancing potential as shown remarkable increase in per cell violaceinprosuction even at lower concentrations. But the highest increase (344.44%) was obtained at 500 µg/ml concentration followed by 750µg/ml.

> ## HPLC of Ethanolic extract of rayan

For reverse phase HPLC (Agilent 1200 infinity) 5µl of ethanolic extract of rayan (reconstituted in ethanol) was injected into phosphoric acid: water (95:5) equilibrated semi-preparative C_{18} column (Eclipse plus C_{18}-Agilent). The column was eluted for 35 min in phosphoric acid: water followed for 35 min with a linear phosphoric acid: acetonitrile (50:50) gradient as mobile phase at the flow rate of 1mL/min. Absorbance was monitored at 220 nm.

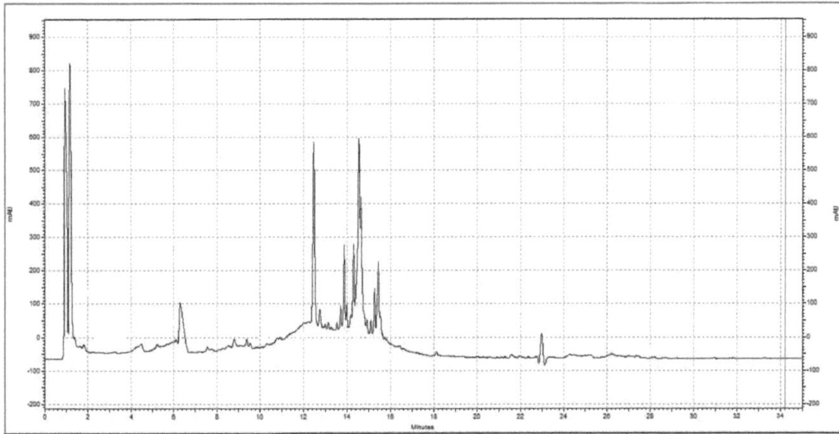

Fig.2. Reverse phase HPLC of Ethanolic extract of rayan

In Figure 2 the compounds are represented by different peaks separated in time in the chromatogram. Each elutes at a specific location, measured by the elapsed time between the moment of injection [time zero] and the time when the peak maximum elutes. By comparing each peak's retention time [t_R] with that of injected reference standards in the same chromatographic system [same mobile and stationary phase], It may be able to identify each compound. Here the polar compounds were eluted first and then after nonpolar compounds were eluted, herepolar compound has the best resolution peak, so we can simply said that more than one polar compound which were present in ethanolic extract of rayan, they were responsible for the quorum sensing enhancing effect against *Chromobacteriumviolaceum.*

Table: 7. Quorum sensing activity of Methanolic effect of rayan

Conc. (µg/mL)	Growth (OD_{660}) (Mean±SD)		%change compared to control	Violacein (OD_{585}) (Mean±SD)		%Change compared to control	Volacein Unit (OD_{585}/OD_{660})		%change compared to control
	C	E		C	E		C	E	
250	1.20±0.02	1.22±0.02	1.66↑	0.36±0.03	0.31±0.01	-13.88	0.30	0.25	-16.66
500	1.20±0.02	1.42±0.04	18.33[*]↑	0.36±0.03	0.42±0.02	16.66↑	0.30	0.29	-3.33
750	1.25±0.01	1.90±0.07	52.00[**]↑	0.26±0.00	0.82±0.08	215.38[*]↑	0.20	0.43	115↑
1000	1.25±0.01	1.94±0.05	55.2[**]↑	0.26±0.00	0.67±0.04	157.69[**]↑	0.20	0.34	70↑

[*]$p<0.05$; [**]$p<0.01$

Here, from the above table it can be seen that methanolic extract of Rayan was found giving different effect when tested at different concentrations. It also enhanced Quorum sensing in *Chromobacteriumviolaceum*. Extarct showed highest quorum sensing activity at the concentration of 750µg/ml.The growth has dose dependent response.

➢ HPLC of Methanolic extract of rayan

For reverse phase HPLC (Agilent 1200 infinity) 5µl of ethanolic extract of rayan (reconstituted in methanol) was injected into phosphoric acid: water (95:5) equilibrated semi-preparative C_{18} column (Eclipse plus C_{18}-Agilent). The column was eluted for 35 min in phosphoric acid: water followed for 35 min with a linear phosphoric acid: acetonitrile (50:50) gradient as mobile phase at the flow rate of 1mL/min. Absorbance was monitored at 220 nm.

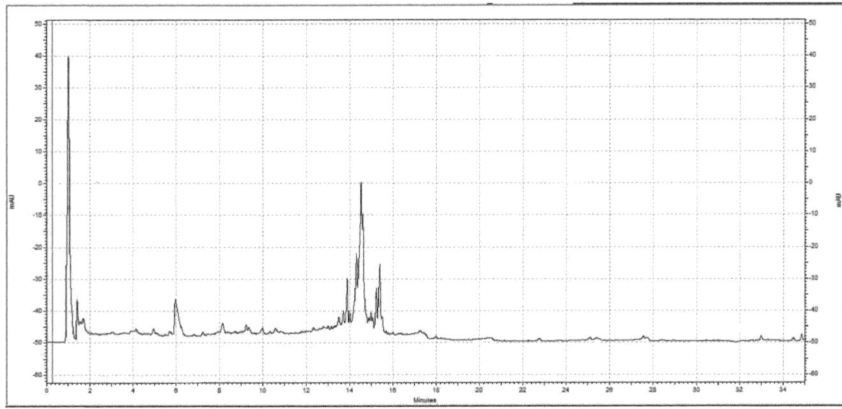

Fig.3. Reverse phase HPLC of Methanolic extract of rayan

In Figure 3 the compounds are represented by different peaks separated in time in the chromatogram. Each elutes at a specific location, measured by the elapsed time between the moment of injection [time zero] and the time when the peak maximum elutes. By comparing each peak's retention time [t_R] with that of injected reference standards in the same chromatographic system [same mobile and stationary phase], It may be able to identify each compound. Here the polar compounds were eluted first and then after nonpolar compounds were eluted, herepolar compound has the best resolution peak, so we can simply said that polar compounds which were present in methanolic extract of rayan, they were responsible for the quorum sensing enhancing effect against *Chromobacteriumviolaceum*.

Table: 8. Quorum sensing activity by Acetone extract of rayan

Conc. (μg/mL)	Growth (OD_{660}) (Mean±SD)		% inhibition compared to control	Violacein (OD_{585}) (Mean±SD)		% inhibition compared to control	Volacein Unit (OD_{585}/OD_{660})		%change compared to control
	C	E		C	E		C	E	
250	1.85±0.03	1.49±0.00	-19.45[**]	0.62±0.02	0.60±0.01	-3.22	0.33	0.40	21.21↑
500		1.62±0.03	-12.43[*]		0.55±0.02	-11.29	0.33	0.33	0
750		1.56±0.01	-15.67[**]		0.58±0.02	-6.45	0.33	0.37	12.12↑
1000		1.49±0.02	-19.45[**]		0.56±0.03	-9.67	0.33	0.37	12.12↑

[*]$p<0.05$; [**]$p<0.01$

Acetone extract of Rayan when tested for quorum sensing showed that growth was affected compared to violacein. Here violacein was inhibited at 500µg/ml. At lower concentration per cell violacein production was obtained as maximum compared to the higher concentrations.

> ➤ **HPLC of Acetone extract of rayan**

For reverse phase HPLC (Agilent 1200 infinity) 5µl of ethanolic extract of rayan (reconstituted in Acetone) was injected into phosphoric acid: water (95:5) equilibrated semi-preparative C_{18} column (Eclipse plus C_{18}-Agilent). The column was eluted for 35 min in phosphoric acid: water followed for 35 min with a linear phosphoric acid: acetonitrile (50:50) gradient as mobile phase at the flow rate of 1mL/min. Absorbance was monitored at 220 nm.

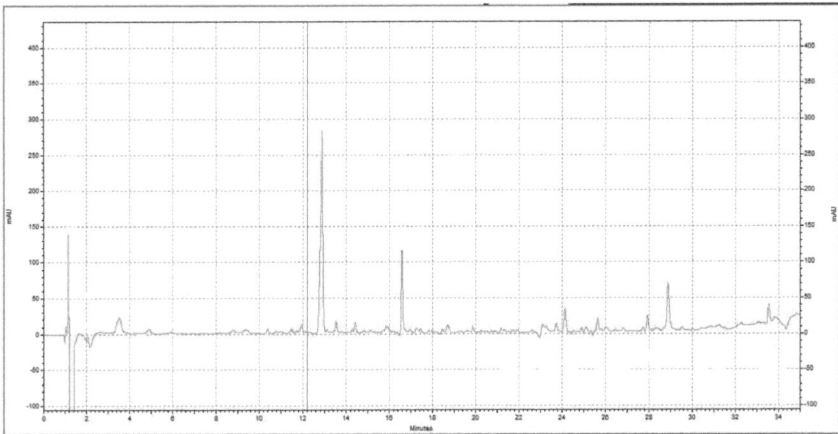

Fig.4. Reverse phase HPLC of Acetone extract of rayan

In Figure 4Very fewer compounds were separated in time in the chromatogram. Each elutes at a specific location, measured by the elapsed time between the moment of injection [time zero] and the time when the peak maximum elutes. Here the polar compounds were eluted first and then after nonpolar compounds were eluted, here polar compound has the best resolution peak, so we can simply said that polar compounds which were present in acetone extract of rayan, they were responsible for inhibition of growth as well violacein in *Chromobacteriumviolaceum*.

Table: 9. Quorum sensing enhancing by methanol nashpati

Conc. (µg/mL)	Growth (OD$_{660}$) (Mean±SD)		%change compared to control	Violacein (OD$_{585}$) (Mean±SD)		% change compared to control	Volacein Unit (OD$_{585}$/OD$_{660}$)		%change compared to control
	C	E		C	E		C	E	
250	1.28±0.02	1.34±0.03	4.68↑	0.18±0.00	0.21±0.01	16.66↑	0.14	0.15	7.14↑
500	1.30±0.00	0.70±0.00	-46.15[**]	0.68±0.00	0.80±0.02	17.64[*]↑	0.52	1.14	119.23↑
750	1.09±0.00	0.49±0.00	-55.04[**]	0.89±0.00	1.03±0.00	15.73[**]↑	0.81	2.10	159.25↑
1000	1.08±0.00	0.46±0.00	-57.40[**]	0.80±0.01	0.94±0.01	17.5[**]↑	0.74	2.04	175.67↑

[*]$p<0.05$; [**]$p<0.01$

Methanolic extract of Nashpati has enhanced quorum sensing in *Chromobacteriumviolaceum* at much higher percentage. Here methenolic extract was precipitated above 250µg/mL.By increasing concentration, the extract enhanced per cell violacein production in *Chromobacteriumviolaceum*. In the case of this extract growth of organism decreased but at the same time violacein production was increased. Here violacein and growth are affected differently.

Other extracts of nashpati were highly precipitated in nutrient broth as well as in mullerhinton media.

5. Effect of different solvents on quorum sensing activity of *C.violaceum*

DMSO, Ethanol and methanol was used as solvent which have antibacterial potential .Here they were tested against organism at different concentration.

Table: 10. Quorum sensing enhancing by DMSO

Conc.	Growth (OD$_{660}$) (Mean±SD)		% Inhibition compared to control	Violacein (OD$_{585}$) (Mean±SD)		% change compared to control	Volacein Unit (OD$_{585}$/OD$_{660}$)		%change compared to control
	C	E		C	E		C	E	
1%	1.04±0.00	1.01±0.01	2.88	0.16±0.00	0.19±0.00	18.75[*]↑	0.15	0.18	20↑
2%	1.04±0.00	0.95±0.01	8.65[*]	0.16±0.00	0.20±0.00	25[**]↑	0.15	0.21	40↑
3%	1.04±0.00	0.87±0.01	16.34[**]	0.16±0.00	0.22±0.00	37.5[**]↑	0.15	0.25	66.66↑
4%	1.33±0.01	1.10±0.00	17.29[**]	0.22±0.01	0.31±0.00	40.90[*]↑	0.16	0.28	75↑
5%	1.04±0.00	0.86±0.01	17.30[**]	0.16±0.00	0.26±0.00	62.5[**]↑	0.15	0.30	100↑

[*]p<0.05; [**]p<0.01

Here DMSO has enhanced quorum sensing in *Chromobacteriumviolaceum* at much higher percentage. By increasing concentration, the DMSO enhanced per cell violacein production in *Chromobacteriumviolaceum*. Lower concentrations of solvents, which apparently do not affect the bacterial growth significantly. In the case of this solvent growth of organism decreased but at the same time violacein production was increase. The production of violacein was dose dependent.

Table: 11. Quorum sensing activity by Methanol

Conc.	Growth (OD$_{660}$) (Mean±SD)		% Inhibition compared to control	Violacein (OD$_{585}$) (Mean±SD)		% change compared to control	Volacein Unit (OD$_{585}$/OD$_{660}$)		%change compared to control
	C	E		C	E		C	E	
1%	1.84±0.01	1.84±0.01	0	0.62±0.04	0.66±0.00	6.45↑	0.33	0.35	6.06↑
2%	1.52±0.02	1.34±0.02	11.84[*]	0.36±0.00	0.35±0.00	-2.77	0.23	0.26	13.04↑
3%	1.52±0.02	1.17±0.03	23.02[**]	0.36±0.00	0.29±0.01	-19.44[*]	0.23	0.24	4.34↑

Conc.	Growth C	Growth E	% inhibition	Violacein C	Violacein E	%inhibition	Volacein Unit C	Volacein Unit E	%change
4%	1.52±0.02	1.01±0.00	33.55**	0.36±0.00	0.23±0.00	-36.11*	0.23	0.22	-4.34
5%	1.28±0.04	0.45±0.00	64.84**	0.23±0.04	0.03±0.00	-86.95*	0.17	0.06	-64.70

*$p<0.05$; **$p<0.01$

Here, from the above table it can be seen that methanol was found giving different effect when tested at different concentrations. It enhanced Quorum sensing in *Chromobacteriumviolaceum*at the level of 3% v/v and also showed highest quorum sensing activity at 5% followed by 4%.

Table: 12. Anti-Quorum sensing activity by Ethanol:

Conc.	Growth (OD_{660}) (Mean±SD)		% inhibition compared to control	Violacein (OD_{585}) (Mean±SD)		%inhibition compared to control	Volacein Unit (OD_{585}/OD_{660})		%change compared to control
	C	E		C	E		C	E	
1%	1.84±0.01	1.60±0.00	13.04**	0.62±0.04	0.58±0.03	6.45	0.33	0.36	9.09↑
2%	1.84±0.01	1.43±0.04	22.28**	0.62±0.04	0.40±0.02	35.48*	0.33	0.27	-18.18
3%	1.84±0.01	1.17±0.04	36.41**	0.62±0.04	0.28±0.01	54.83*	0.33	0.23	-30.30
4%	1.84±0.01	0.20±0.01	89.13**	0.62±0.04	0.03±0.00	95.16**	0.33	0.15	-54.54
5%	1.52±0.02	0.02±0.00	98.68**	0.36±0.00	0	100**	0.23	0	-100

*$p<0.05$; **$p<0.01$

Ethanol had shownthe highest anti-quorum sensing activity at highest concentration. Ethanol scored better followed by methanol and dmso, in terms of their anti-quorum sensing determination. In the case of solvents specifically in ethanol, at highest concentration growth of organism decreased and at the same time violacein production was also decreased. Here inhibition of growth and violacein were signifiacanty increased as concentration was increased. Here growth and violacein were dose dependent at significant level.

4. Discussion

The aim of this study was to determine the anti-QS potential as well as quorum sensing enhancing potential of medicinal plants as a means to uncover a potential mode of action and to suggest a validation for continued traditional use. To date, research has shown anti-QS in only a few higher plants and seaweed (Manefieldet al., 1999; Teplitski et al., 2000; Fray, 2002; Gao et al., 2003). This study introduces three medicinal plant species in which one having anti-QS activity and other two having quorum sensing enhancing activity (Table: 1).

In view of limited supply of natural fossil fuel, *Pongamia* is undoubtedly one of the key source species and a potential source of viable biodiesel. The study reported here has an applied significance.
It is clear that *pongamia* has antibacterial compound against enteric pathogens and that can be used as a treatment of enteric infectious (S R Arote et al., 2009; VigyaKesari et al., 2009). *pongamia* has also anti quorum sensing activity .it shows best anti-quorum sensing activity in ethanolic and methanolic extract compared to aceton extract.

In *M.hexandra* the presence of antibacterial and antifungal substances is higher and well established. In *M.hexandra* plant mainly the leaf portion is used for the therapeutic purpose.It has potential to enhancing quorum sensing.*P.pyrifolia* has also capacity to increase quorum sensing.extracts which have capacity to increase the violacein production they are very important because violacein exhibits important antitumoural, antimicrobial and antiparasitary properties (Nelson Durán., et al 2010).

In summary, the anti-QS potential of medicinal plants may be as important as the antibacterial effect. In this study, we have shown three medicinal plants to have QS activity. We propose that the QS effect of the plants reported could contribute in part to their efficacy and traditional use as medicines. Although biomonitor strains were chosen for ease of screening, similar QS systems exist in many human pathogens.

References:

1. Shahid M, Shahzad A, Sobia F, Sahai A, Tripathi T, Singh A, et al. Plant natural products as a potential source for antibacterial agents: Recent trends. Anti-infect Agents Med Chem2009; 8: 211-225.
2. Rasmussen, T.B., Bjarnsholt, T., Skindersoe, M.E., Hentzer, M., Kristoffersen, P., Kote, M., Nielsen, J., Eberl, L., Givskov, M., 2005a. Screening for quorum-sensing inhibitors (QSI) by use of a novel genetic system, the QSI selector. J. Bacteriol. 187, 1799–1814.
3. Groves M. G., J. M. Strauss, J. Abbas, and C. E. Davis. 1969. Natural infections of gibbons with a bacterium producing violet pigment (*Chromobacterium violaceum*). J. Infect. Dis. 120:605–610
4. Gordon MC, David JN. Natural product drug discovery in the next millennium. Pharm Biol. 2001; 39: 8–17.
5. Gao M., Teplitski M., Robinson J.B., Bauer, W.D., 2003. Production of substances by *Medicagotruncatula* that affect bacterial quorum sensing. The American Phytopathological Society 16, 827–834.
6. Duran, N., and C. F. Menck. 2001. *Chromobacterium violaceum*: a review of pharmacological and industrial perspectives. Crit. Rev. Microbiol. 27:201